CHEMISTRY
Laboratory Notebook

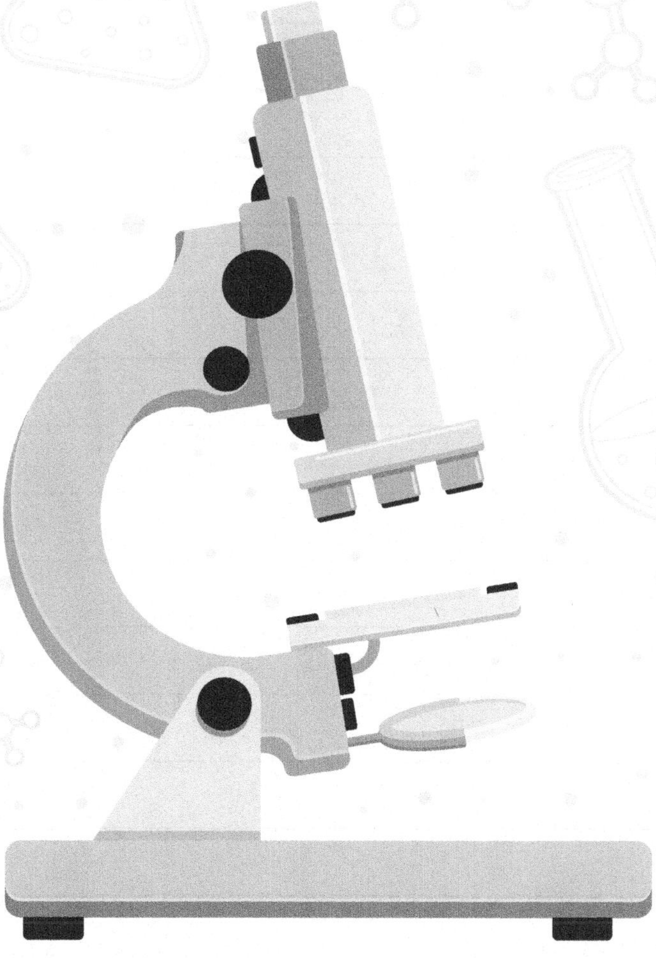

Table of Contents

No.	Date	Subject / Project / Experiment	Page

Table of Contents

No.	Date	Subject / Project / Experiment	Page

Table of Contents

No.	Date	Subject / Project / Experiment	Page

Table of Contents

Title

Name

Date

Signature

Date

Signature

Date

Title

Name

Date

Signature

Date

Signature

Date

Title

Name

Date

Signature

Date

Signature

Date

Title

Name

Date

Title

Name

Date

<table>
<tr><td>Subject / Project / Experiment</td><td>Page</td></tr>
<tr><td>Title</td><td>Name</td><td>Date</td></tr>
<tr><td>Signature</td><td>Date</td><td>Signature</td><td>Date</td></tr>
</table>

Title

Name

Date

Signature

Date

Signature

Date

Title

Name

Date

Title

Name

Date

Title

Name

Date

Subject / Project / Experiment

Page

Title

Name

Date

Signature

Date

Signature

Date

Subject / Project / Experiment

Page

Title

Name

Date

Signature

Date

Signature

Date

Title

Name

Date

Subject / Project / Experiment

Page

Title

Name

Date

Signature

Date

Signature

Date

Title

Name

Date

Signature

Date

Signature

Date

Title

Name

Date

Signature

Date

Signature

Date

Title

Name

Date

Signature

Date

Signature

Date

Title

Name

Date

Signature

Date

Signature

Date

Title

Name

Date

Signature

Date

Signature

Date

Title

Name

Date

Title

Name

Date

Signature

Date

Signature

Date

Title

Name

Date

Title

Name

Date

Signature

Date

Signature

Date

Title

Name

Date

<table>
<tr><td>Subject / Project / Experiment</td><td>Page</td></tr>
<tr><td>Title</td><td>Name</td><td>Date</td></tr>
</table>

<table>
<tr><td>Signature</td><td>Date</td><td>Signature</td><td>Date</td></tr>
</table>

Subject / Project / Experiment

Page

Title

Name

Date

Signature

Date

Signature

Date

<table>
<tr><td>Subject / Project / Experiment</td><td>Page</td></tr>
<tr><td>Title</td><td>Name</td><td>Date</td></tr>
</table>

Signature

Date

Signature

Date

Title

Name

Date

Signature

Date

Signature

Date

Title

Name

Date

Signature

Date

Signature

Date

Title

Name

Date

Title

Name

Date

Title

Name

Date

Signature

Date

Signature

Date

Title

Name

Date

Title

Name

Date

Title

Name

Date

Title

Name

Date

Signature

Date

Signature

Date

<table>
<tr><td colspan="2">Subject / Project / Experiment</td><td>Page</td></tr>
<tr><td>Title</td><td>Name</td><td>Date</td></tr>
</table>

<table>
<tr><td>Signature</td><td>Date</td><td>Signature</td><td>Date</td></tr>
</table>

Title

Name

Date

Subject / Project / Experiment

Page

Title

Name

Date

Signature

Date

Signature

Date

Title

Name

Date

Signature

Date

Signature

Date

Title

Name

Date

Title

Name

Date

Title

Name

Date

Made in the USA
Monee, IL
07 July 2026

56551365R00063